Filipe Aguiar e Silva
Cecília R. M. e Silva
Eloísa O. S. Saliba

Techniques for assessing consumption and digestibility in ruminants

Filipe Aguiar e Silva
Cecília R. M. e Silva
Eloísa O. S. Saliba

Techniques for assessing consumption and digestibility in ruminants

Using internal and external indicators to assess digestive parameters in cattle

ScienciaScripts

Imprint

Cover image: www.ingimage.com

This book is a translation from the original published under ISBN 978-613-9-62676-2.

Publisher:
Sciencia Scripts
is a trademark of
Dodo Books Indian Ocean Ltd. and OmniScriptum S.R.L publishing group

120 High Road, East Finchley, London, N2 9ED, United Kingdom
Str. Armeneasca 28/1, office 1, Chisinau MD-2012, Republic of Moldova, Europe
Printed at: see last page
ISBN: 978-620-7-74900-3

SUMMARY

ACKNOWLEDGMENTS 2
SUMMARY 3
CHAPTER 1 - GENERAL INTRODUCTION 4
CHAPTER 2- LITERATURE REVIEW 6
Chapter 3 - USE OF DIFFERENT INDICATORS IN DETERMINING CONSUMPTION, DIGESTIBILITY AND FECAL PRODUCTION IN BOViNs, COMPARED WITH TOTAL FECAL COLLECTION 22

ACKNOWLEDGMENTS

To GOD, through him we were able to achieve all our goals and find the possibility where we thought it was impossible.

To my parents Heloisa (*in memorian*) and Carlos Alberto, my brother Lucas and my girlfriend Cecilia for their trust, encouragement and understanding.

To my advisor, Dr.ª Eloisa de Oliveira Simoes Saliba, for the opportunity, trust, patience and teachings she placed in me.

To Professor Dr. Geraldo Sérgio Senra Carneiro Barbosa for his support, trust and for being present at every decision.

To Professor Dr. Norberto Màrio Rodriguez for being part of this achievement.

To Dr. Guilherme Rocha Moreira, Cristovao Colombo de Carvalho Couto Filho, and MSc. Nélio Cunha Gonçalves for their great help in carrying out this study.

To Kelly, Marcos and Toninho for their help during the laboratory period.

To CNPQ, INCT and CAPES for their support in carrying out the projects.

Finally, to everyone who contributed to this work.

SUMMARY

Techniques for estimating consumption, digestibility and fecal production with different diets for cattle

The aim of this study was to evaluate the internal indicators, MSi and FDNi at 264 hours of *in situ* incubation, MSi and FDNi at 72 and 144 hours of *in vitro* incubation, and the external indicators LIPE® and TiO_2 with total collection in determining digestibility, consumption and fecal production of Dutch x Gir crossbred heifers on three different diets: corn silage + concentrate (corn and soy), sugar cane + concentrate and tifton hay + concentrate. The experimental design used was a triple and simultaneous 3 x 3 Latin square. The results obtained show that only the MSi and FDNi incubated *in vitro* for 72 hours did not provide statistically similar results to the total collection. The other indicators were able to provide statistically similar estimates to the total collection when evaluating consumption, digestibility and fecal production.

Keywords: heifers, nutrition, roughage.

CHAPTER 1 - GENERAL INTRODUCTION

1 INTRODUCTION

Determining the nutritional value of food has been the aim of several studies involving animals. However, evaluating these foods is a challenge for nutritionists. In this context, the study of digestibility, consumption and fecal production is an essential practice in the formulation of animal diets.

Digestibility and consumption are recognized as the main determinants of the quality of a feed (Rodriguez et al., 2006). They are therefore limiting factors for the animal's productive efficiency, depending on the characteristics of the food, the environment and its interactions, and the inherent factors of the animal itself.

In addition, during the process of passing food through the gastrointestinal tract, the digestibility of the diet generates a potentially indigestible residue, which is not used by the animal and is determined as fecal production.

The points formed by combining knowledge of fecal production, animal consumption and the digestibility of the diet offered, allow the nutritionist to formulate diets that can possibly be better used by the animals.

However, direct determination of fecal production, known as *in vivo* total fecal collection, is a costly process, requiring control of daily intake and excretion, adaptation of animals to cages or stalls, or to collection bags in grazing animals, which may not be feasible in some situations.

Although *in vivo* procedures have their limitations in some cases, they remain the benchmark for both food evaluation and the validation of alternative determination methods. In order to develop alternatives that make it possible to estimate fecal production, digestibility, and indirectly consumption, indicator techniques have been proposed, which are chemical monitors according to Owens and Hanson (1992) for quantitative and qualitative determination of physiological and nutritional phenomena.

Determining digestibility using indicators does not require manipulation of total fecal output and strict control of actual consumption, as the known fractions of the substance used as an indicator in the feces and the dose administered are taken into account.

Since then, several experiments have been carried out in order to assess the dynamics of the indicators in the animal digestion process, i.e. whether there are interactions between them and the animal's digestive tract or with the diet offered; whether the delivery method or laboratory evaluation techniques have an effect on the recovery of their fractions in the fecal bolus; and whether these sets of information can reliably represent fecal production, consumption and digestibility.

The aim of this study was to study and determine the best technique for these estimates. For this purpose, the determination through total collection was evaluated as a reference for the indicators indigestible dry matter, indigestible neutral detergent fiber, titanium dioxide and enriched purified lignin.

CHAPTER 2- LITERATURE REVIEW

1. INTERACTION BETWEEN CONSUMPTION, DIGESTIBILITY AND FECAL PRODUCTION

Consumption, digestibility and fecal production are the three pillars of studies that seek to understand the quality of animal feed. Consumption and digestibility are direct indicators of the nutritional value of a diet, while fecal production is a useful tool for estimating the first two.

Knowledge of two of these factors will always allow the third to be estimated. Following the relationship: the production of fecal dry matter (FP) is equal to the dry matter intake (DMI) multiplied by the fraction of the feed that is not absorbed, i.e. one minus the dry matter digestibility coefficient (DMS). This can be represented mathematically as follows:

PF = CMS x (1 - DMS)

or

CMS = PF /(1 - DMS)

or

DMS = 1 - (PF /CMS)

The most precise methodology, which serves as a parameter for measuring these variables, would be to weigh the feed and feces directly to obtain consumption and fecal production, followed by the application of the mathematical formula expressed above to obtain digestibility. However, in studies with cattle, where a larger number of animals are usually worked with and in environments that are difficult to control, such as grazing, this practice becomes difficult to apply. For this reason, simpler techniques have been proposed for determining these variables.

1.1 METHODOLOGIES FOR MEASURING CONSUMPTION, DIGESTIBILITY AND FECAL PRODUCTION

Determining the quality and use of forage by the animal involves numerous barriers, especially when the animal is grazing. Information on the consumption of forage under grazing is restricted to specific conditions and limited measurement procedures, which makes it impossible to use it in a more practical way.

According to Astigarraga (1997), estimating the consumption of animals managed under grazing conditions is so complex that all the methods used have limitations and compromises that can lead to errors.

Le Du & Penning (1982) reported that consumption estimates should be obtained using techniques that cause the animal to deviate as little as possible from its grazing activities.

Various methodologies for estimating the consumption of grazing animals have been published. Minson (1990) proposed classifying these methods according to the length of the estimation period. Burns et al. (1994) proposed a grouping of these methods, using a direct, indirect or empirical estimate of consumption. Moore & Sollenberger (1997) considered all consumption estimates for grazing animals to be indirect measurements, classifying them as individual consumption estimates, or consumption estimates for groups of animals, or for the pasture. Greenhalgh (1982) reported that methods for estimating forage consumption under grazing conditions can be divided into two broad categories: those based on pasture disappearance, and those based on animal gain.

Mayes (1989) concluded that estimates of consumption at pasture can be obtained through methodologies based on measurements made in the pasture (differences in pasture weight), or variables observed in the animal itself (difference in animal weights, ingestive behavior, animal performance, fecal production and digestibility).

When confined animals are used, whether in cages or even stalls, measurements are more precise and concise, as there is strict control of the diet offered to the animals, the leftovers found in the post-feeding trough, and fecal production.

From the strict control of observations in confined animals and the consequent validation of the different techniques, the methodologies can be extrapolated to grazing animals. The use of fecal production and digestibility indicators is undoubtedly a clear example of the commitment of researchers who, *in* various works, compare *in vivo techniques* (direct) with indirect techniques (indicators) in confined animals, for later use in grazing animals.

1.2 PRINCIPLES OF USING THE INDICATOR TECHNIQUE

In many cases, faced with the numerous difficulties or impossibilities of carrying out total feces collection and controlling dietary intake (direct methodology), researchers have used inert substances known as indicators to determine digestibility, fecal production and consumption. These compounds are simple to manipulate and have the potential to provide information similar to that from *in vivo* tests.

According to Owens & Hanson (1992), indicators are compounds used to monitor chemical (hydrolysis and synthesis) and physical (flow) aspects of digestion and are routinely used to estimate digesta flow and fecal production in ruminants.

Saliba (1998) reports that an ideal indicator must have some fundamental properties, the main ones being that it is not toxic, has no physiological function, is not metabolized, is completely recoverable from the digestive tract and has no influence on intestinal motility and secretions.

An ideal indicator for estimating fecal production may not be suitable for estimating kinetics, due to particle migration, phase separations, inhibition of digestion, osmotic effect in the intestine, among others, according to Rodriguez et al. (2006).

The use of fecal indicators is based on the fact that as the feed follows the flow of the digestive tract, the concentration of the indicator progressively increases due to the removal of digestible constituents from the feed ingested by the animal, and that these concentrations of non-absorbable substances can be recovered in the feces. The increase in concentration is proportional to digestibility and the latter can therefore be calculated from the concentrations of the indicator in the feed.

Incomplete recovery in feces, variations in rumen flow, unrepresentative sampling and statistical design are primary problems associated with experiments using indicators (Merchen, 1988; Titgemeyer, 1997).

To determine fecal output, once the above has been taken into account, it is enough to use a simple mathematical relationship: fecal output (FP) is equal to the ratio between the amount of indicator ingested and the concentration of the indicator in the feces:

$$PF = \frac{\text{Ingested dose}}{\text{Fecal concentration}}$$

Indicators can be classified as internal, represented by indigestible substances naturally present in some component of the diet, and external, when added to the diet or supplied orally (Berchielli et al., 2005).

The biggest obstacle to using internal indicators is their variable recovery in the feces. For external indicators, it is the fact that some of them can aggregate with the fibrous fractions of the food, which can alter its chemical and physical fractions (Rodriguez et al., 2006).

The known internal indicators, which are natural constituents of diets and are neither digested nor absorbed by animals, are: Silica; Fecal Nitrogen; Chromogen; Lignin; indigestible neutral detergent fiber (NDF) and acid detergent fiber (FDA), Acid Insoluble Ash (ASI) and Acid Detergent Insoluble Ash (ADI); and N-alkanes, the last four being the most commonly used.

The following are known as external indicators: chromic oxide, the rare earth elements (lanthanum, samarium, cerium, ytterbium, dysprosuim), ytterbium being the most widely used of these elements, mordant chromium and LIPE®, which are used for the solid phase. For the liquid phase, Cobalt E.D.T.A., Chromium E.D.T.A. and Polyethylene Glycol (PEG) are used (Rodriguez et al., 2006).

For the use of indicators to be validated, they must always be compared to a standard, total stool collection. In addition, each researcher must standardize the use of the indicator for certain situations, aware of possible variations and errors in sampling. The choice of indicator is determined by the objective of the research in question, taking into account costs, ease of handling and analysis.

1.2.1INDICATORS

1.2.1.1 PURIFIED AND ENRICHED LIGNIN (LIPE)®

The modified and enriched hydroxyphenylpropane, LIPE® , is an external digestibility indicator that was developed specifically for research.

The first evaluations using LIPE® to estimate intake and digestibility compared to total fecal collection were carried out on rabbits. Other estimates of fecal production and digestibility revealed the efficiency of LIPE® as an external indicator, showing accuracy in relation to the data obtained with total collection. Among other advantages, this indicator provided a short adaptation period, as well as being low cost.

Saliba et al. (2003) in a trial evaluating Tifton 85 hay in diets for sheep, found a similar average digestibility coefficient of 63.23% and 64.78%, and fecal production of 365.39 g/day and 383.07 g/day, for LIPE® and total collection, respectively.

Lima et al, (2008) in a trial testing chromic oxide and LIPE® for estimating dry matter intake in beef cattle, recommended LIPE® as the most reliable option for indirectly determining dry matter intake in animals grazing on pasture. Corroborating this data, Ferreira et al. (2009) evaluated indicators in cattle for estimating digestibility and found that LIPE® accurately estimated digestibility, similar to the total feces collection method.

Moraes (2007), working with estimates of fecal production in goats fed different agro-industrial by-products using the LIPE® , concluded that this indicator did not differ from the total fecal collection method, providing accurate measurements in estimating the apparent digestibility of nutrients.

Therefore, most of the studies in which LIPE® has been evaluated for estimates of fecal production, digestibility and consumption in different species have shown it to be effective, economically viable and easy to handle and determine.

The adaptation period for uniform excretion is only 48 hours and the period for collecting feces is three days for birds and five days for other species. The analytical technique used to measure the indicator in feces is near-infrared spectroscopy, which is quick, sensitive and non-destructive to the sample.

1.2.1.2 TITANIUM DIOXIDE (THIO2)

Titanium dioxide TiO_2 is a white powder with no smell or taste, and has emerged as an alternative to chromic oxide (Cr O_{23}) as it has no carcinogenic properties and has been used as an indicator in digestibility studies in cattle, pigs, poultry and rats (Myers et al., 2006). Another advantage of titanium dioxide over chromic oxide is that it is approved as a dietary additive by the Food and Drug Administration (USA) (Lopes, 2007). TiO2 is commonly found in some human food products, used as a coloring agent and as part of the coating composition of chewing gums. In addition, it is used as a pharmaceutical additive where it is part of the composition of some sunscreens. Titanium dioxide exhibits great semiconductor absorption of ultra violet (UV) radiation and reflects and scatters radiation in the visible and UV regions. For this reason, they are considered to be the most efficient physical filters, justifying their use in solar preparations (Salgado; Galante; Leonardi, 2004).

Sampaio et al. (2011) state that TiO_2 has no limitations in terms of its inclusion in animal diets. Furthermore, according to Valadares Filho et al. (2006), another competitive advantage of TiO2 is its low cost.

Pina (2008) also found good estimates of concentrate intake and digestibility using the indicators chromic oxide and titanium dioxide, when he worked with Nelore heifers fed hydrolyzed sugar cane with different levels of lime.

Glindemann et al. (2009) evaluated the TiO2 indicator in adult sheep fed hay supplemented or not with concentrate and reported that the average TiO2 recovery was 1.04 kg. TiO2 recovery was higher ($P < 0.001$) in hay diets than in hay and concentrate diets (1.08 and 0.99 kg, respectively).

One of the main problems with determining titanium in stool samples is choosing a technique that really gives accurate results of the amount of TIO2 found in the sample. There may be qualitative techniques used for its determination, but they fall short when it comes to the quantities found in certain samples. These techniques only indicate the presence of the compound.

1.2.1.3 indigestible food fractions

When using an indigestible component present in the feed as an indicator, its concentration in the feces is a function of the different digestive events to which the digesta is subjected, thus establishing a cause/effect relationship between the feed and the gastrointestinal tract, consistent with the *in vivo* environment (Detmann et al., 2001). Thus, internal indicators have the advantage of already being present in the feed and, in general, remaining evenly distributed in the digesta, which gives them advantages over external indicators (Piaggio et al., 1991).

The main internal indicators reported in the literature are acid insoluble ash (ASI), acid detergent insoluble ash (ADI), lignin, cutin and n-alkanes (Lopes, 2007). However, the use of some of these indicators has been limited by their low concentrations in the food - leading to an increase in the standard deviation - among other problems, such as contamination of the food, feces and leftovers with sand and the difficulties imposed by the analytical methods used.

In order to minimize sampling errors due to the low concentration of some internal indicators, indigestible fibrous fractions of the feed, obtained after incubations (*in vitro* or *in situ*), were suggested by Lippke *et al.* (1986), and have been evaluated by several authors (Cochran et al., 1986; Berchielli et al., 2000; Valadares Filho, 2000; itavo et al., 2002; Zeoula et al., 2002; Brito et al., 2007; Detmann et al., 2007; Cabral et al., 2008).

Indigestible neutral detergent fiber (FDNi), indigestible acid detergent fiber (FDAi), indigestible dry matter (MSi), acid detergent lignin (LDA) and potentially indigestible cellulose (CELi) are examples of these indicators, of which the first three stand out as the most widely used.

However, the use of indicators derived from this methodology is not an easy task, as there is no standard method for analyzing indigestible fractions, and the faecal recovery of these indicators has been widely variable, which compromises the precision and accuracy of digestibility estimates (Penning & Johnson, 1983; Cochran et al., 1986; Lippke et al., 1986; Berchielli et al., 2000).

When choosing to use indigestible fiber as an internal indicator, some factors of variation must be considered, such as the composition of the dietary fiber, the size of the incubated particles, the number of days and times of collection of the feces, the period of time and the mode of incubation (*in vitro* or *in situ*).

Some studies have shown in their results that there may be an interaction between the indicator and the type of fiber used as bulk in animal trials (Zeoula et al., 2002; Berchielli et al., 2005).

To obtain the undigested fraction of a given feed, a prolonged incubation time is necessary. According to Berchielli et al. (2000), there is a reduction in the digestibility of feed when the exposure time of the sample to the rumen microbiota is reduced, which also leads to an unrealistic reproduction of the undigestible fraction of the samples. In studies with internal indicators, this time, whether *in vitro* or *in situ* incubations, has varied greatly.

Some authors have reported that unevenness in the size of the particles incubated can be the cause of incomplete fecal recovery of internal indicators (Lippke et al., 1986; Fondevila et al., 1995). In this context, both the size of the particles to be incubated and the porosity of the nylon bags are important.

The porosity, type and size of the filter bags used for *in situ* incubation, as well as the amount of sample incubated, is another source of variation in research using internal indicators. Some authors choose to use standardized filter bags marketed for this purpose, such as those associated with the Ankom® system, which are containers made of synthetic fibre insoluble in acid and neutral media and called "Filter Bags", F57 (Alves et al., 2003; Detmann et al., 2007; Dias et al., 2007.).

Considering all the points discussed in the previous paragraphs, there is a lack of uniformity in the methodological procedures used to obtain indigestible residues, making it necessary to standardize the techniques in order to obtain more reliable results.

1.3 USE OF PRESERVED FODDER FOR ANIMAL FEED

The seasonality of forage production is one of the main factors responsible for the low productivity rates of the country's livestock, since the levels of animal production obtained during the rainy season are compromised by low forage yields during the dry season. In Brazil as a whole, there are different rainy seasons throughout the year, due to the country's size. In the southeastern region of the country, the period from May to September is the time when there is a marked reduction in the growth of tropical forage grasses, resulting in less availability of forage, which has a low nutritional value, thus affecting the performance of animals kept under grazing conditions.

There are countless strategies that have the potential to overcome forage deficits during the dry period of the year in the different regions of Brazil. However, all the alternatives that can be used must be consistent with each property's production system. The use of conserved fodder, winter fodder, weeding, concentrated supplementation in the pasture and the use of irrigation are alternatives that can exploit the low productivity of tropical pastures during this period of the year.

Preserved fodder provides high quality food more evenly throughout the supplementation period. Silage and hay are the main ways of using preserved fodder.

Silage is green fodder, with varying moisture content between species, stored in anaerobic environments called silos. It is preserved through fermentation.

The paralysis of respiration due to the absence of air and the inhibition of fermentation due to the reduction in pH are the two main immediate events in ensiled fodder. Bacteria act on soluble carbohydrates, transforming them into acids, especially lactic and acetic, which reduce the pH and prevent undesirable fermentation. The material should be ensiled at a pH between 5.5 and 6.0, and preservation and stabilization occurs at pH 3.5 to 4.2.

Haymaking consists of dehydrating green fodder with 65 - 85% moisture to 10 - 20%. The most intense water loss occurs shortly after cutting and decreases as the moisture content drops below 65%. The process continues until minimum moisture values are reached. The optimum haying point occurs in association with the speed with which dehydration takes place. Therefore, from harvesting to the end of the process, certain factors such as the time of year for haying in each region, turning the material in the field after cutting, the permanence of the material in the field and the storage of the final product must be taken into account.

The ideal hay should have a greenish color, a pleasant smell, have a good percentage of leaves, be soft, free of impurities and toxic elements and, above all, have good digestibility.

As for the animal's use of these preserved forages, consumption is a limiting factor. Minson & Wilson (1994) proposed a series of characteristics related to forage intake that are likely to favor consumption or not by animals, such as: chemical and bromatological, physical-anatomical characteristics; and digestive kinetics. All these aspects influence the nutritional value of these forages and consequently the animal's voluntary consumption.

1.4.1CORN SILAGE

Corn production, together with soybeans, accounts for approximately 80% of grain production. In cattle feed, corn is used as a source of starch, which is generally the main energy component of concentrates. In addition to its use as an energy ingredient in concentrates, the maize plant as a whole is used as one of the main bulks preserved in the form of silage for feeding cattle.

For a long time, breeders' objectives were limited to agronomic criteria for corn production, such as grain yield per hectare, resistance to pests and diseases, and climatic stresses (Michael-Doreau and Doreau, 1999). However, the use of maize in the form of whole plant silage has led some companies to work on their hybrids for silage production, even before starting the search for exclusive genotypes (Aseltine, 1988).

Some undesirable aspects for silage production were observed in the graniferous hybrids, such as: early hardening of the grains and high cell wall constituents, which would result in a food with low nutritional value when it comes to silage.

As a result, new hybrids emerged that were suitable for the ensiling process and carried with them better fiber digestibility, a greater number of leaves above the ear line and grains that were better suited to starch utilization.

Some characteristics are of fundamental importance in determining the quality of corn silage for animals. According to Phipps and Weller (1979), silages with a high grain content (50% grain in dry matter) had less acid detergent fiber (FDA), cellulose, lignin and more starch than silages with a low grain content (26% grain in dry matter).

Hemken et al. (1971) concluded that there was little or no advantage of silages with a high grain content on the performance of dairy cows. In view of the minor importance of grain content on cow performance, the same authors recommend planting maize varieties based on dry matter production/ha and not on grain content.

Fiber digestibility is also an important tool for controlling the quality of corn silage. According to Oba and Allen (1999), hybrids with lower lignin concentrations have higher *in vitro* dry matter digestibility and higher dry matter and NDF degradability in the rumen (Bal et al., 2000).

In order to develop more digestible maize hybrids, the Brown Midrib (BM) was discovered. This genetic modification promotes a decrease in the lignin content of the plant, thus favoring its utilization.

Despite their favorable characteristics in terms of digestibility and animal utilization, maize with BM genes is rarely used for silage production due to its susceptibility to diseases, lodging and lower yields per hectare (Gallais et al., 1980).

The biochemical composition of the starch in the grain, the texture, the oil content and the proportion of leaves in relation to the whole plant summarize the rest of the essential characteristics for producing quality corn silage.

The starch content and the amylose:amylopectin ratio vary between the different maize genotypes. In a study carried out by Phillippeau and Michael-Doreau (1997), the values found ranged from 61.0 to 68.6% starch for kernels with soft endosperm and 58.6 to 67.9% for kernels with hard endosperm.

Phillippeau et al (1998) compared the degradability of maize grains from three vitreous and floury genotypes and concluded that the degradation rates of floury grains were higher than those of hard grains.

The leafier hybrids are sold on the market as being specifically designed for whole plant silage production. They are characterized by a greater number of leaves above the ear line and higher moisture contents in the grains and in the whole plant at maturity (Bal et al., 2000).

1.4.2TIFTON 85 HAY

Cynodon plants have characteristics that make them suitable for hay production. They have suitable morphology, especially thin stems and leaves that are well attached to the stalk (Haddad and Castro, 1998). Tifton 85 grass has a high potential for dry matter (DM) production and high digestibility coefficients (Hill et al., 1998).

The Tifton 85 hybrid is the result of crossing the Tifton 68 grass (Cynodon nlemfuensis Vanderyst) with the Bermuda Grass from South Africa, registration PI 290884 Cynodon dactylon (L.) Pers. It is a perennial and stoloniferous hybrid, with a large leaf mass and thick rhizomes, which are underground stems responsible for maintaining the reserves of carbohydrates and nutrients that provide its incredible resistance to droughts, frosts, fires and low grazing.

It was developed by the US Department of Agriculture in cooperation with the University of Georgia, at the Tifton experimental station, by Professor Dr. Burton's team, and was registered and released for planting on 31/10/92. Introduced to Brazil in 1993, it has resisted and remained green in the face of the prolonged frosts and droughts of 1994.

Tifton-85 can be planted in cold regions as well as warm regions with subtropical and tropical climates, i.e. throughout the country, in sandy, mixed and clay soils (not flooded), properly corrected and fertilized as advised by the regional agronomist.

In Brazil, Ribeiro et al. (1998a), evaluating the DM production of Tifton 85 grass at three cutting frequencies, under five nitrogen applications (0, 100, 200, 300 and 400 kg/ha), estimated that, for the application of 300 kg of N /ha, DM yields of 16.67; 19.06; and 21.44 t/ha, respectively, for 28, 42 and 56 days of regrowth.

Highly recommended for its green mass production, high leaf/stem ratio, great resistance and high nutritional value, Tifton - 85 has proven to be an excellent option for green and conserved pastures for cattle.

1.4.3SUGAR CANE

Sugarcane (*Saccharum spp.*) was probably one of the first plants used as a forage resource for supplementary animal feed in Brazil and, due to its productive characteristics and numerous ways of using it, it has acquired significant importance in animal feed. It is a perennial grass native to tropical and subtropical climates and probably originated in Southeast Asia. There are several species of the *Saccharum* genus (*S. officinarum, S. spontaneum, S. robustum,* etc.) but the varieties used today are almost all hybrids.

Its use as a supplementary roughage during the dry season is justified by its ease of cultivation, high productivity, low cost per unit of dry matter produced, as well as its peak production and nutritional value coinciding with the period when forage is scarce.

Another aspect of great importance is that its bromatological composition remains practically unchanged after a certain stage of growth, a fact that places it in an outstanding position in relation to other forage crops, whose percentage of fibrous fractions increases as physiological age advances, while at the same time there is a reduction in their protein content.

When compared to other grasses, such as elephant grass and corn silage, the NDF content of sugarcane is relatively low, at 55.87% for sugarcane and 76.93% and 55.26% for elephant grass and corn silage, respectively. However, the digestibility coefficients of the fibrous fraction of sugarcane are low, which corroborates the fact that it is harvested during the dry season of the year.

Fernandes et al. (2003), considering intervals of four months between cuts of sugarcane, found that the differences in the levels of NDF and acid detergent insoluble fiber (FDA) were relatively small, which shows the ability of this roughage to maintain its nutritional value constant over time, contrary to what happens with most tropical forage species.

The major problem with the profile of the fibrous fraction of sugar cane is its low quality or digestibility, which can limit dry matter intake and the performance of animals fed diets containing sugar cane.

The structural carbohydrates in sugar cane are a potential source of low-cost energy for ruminants. However, its potential as an energy source is limited due to its low digestibility and degradation rate.

In addition to the low quality of the fiber, there are other problems with using sugar cane to feed cattle, such as the low protein and mineral content.

The most promising varieties for feeding cattle are those with the lowest levels of NDF, the highest values of *in vitro* digestibility of dry matter (DIVMS), NDF/Pol ratios (Pol - sucrose content) of less than 2.7 and low levels of lignin. Considering that the species has a low nitrogen content, the CP content is not a criterion for choosing varieties (Costa et al., 2003; Rodrigues et al., 2005, 2006).

Rodrigues et al. (2006) found differences between the varieties studied in terms of lignin content, which ranged from 2.9 to 4.1% of DM. This variable, which is part of the NDF, has a high negative correlation with digestibility, in addition to the fact that an increase in the NDF content in the plant is associated with a thickening of the cell wall, which reduces the area available for microbial attack in the rumen.

Another characteristic that should be evaluated when choosing varieties is the greater capacity for natural or easy defoliation, as this allows for greater efficiency in the cutting and milling process, as well as reducing the supply of material of low nutritional value to the herd (Macedo et al., 2006).

There are studies that show that treating fibrous materials with alkali, such as micropulverized lime, increases their digestibility. The phenomenon most associated with the alkaline treatment of roughage is the partial solubilization of hemicelluloses, lignin and silica, and the hydrolysis of uronic and acetic acid esters. Treatment with alkali can also lead to the breaking of hydrogen bridges in the cellulose. In other words, bonds are broken in the fibrous fraction of the cane, which leads to an increase in its digestibility. The justification for using alkali lies in the fact that the lignin in grasses is particularly susceptible to hydrolytic attack by this base on the ester-type covalent bonds between the lignin and the cell wall (Van Soest, 1994).

In order to decide when to use sugar cane diets, it is necessary to evaluate dry matter production per hectare, animal performance, animal category, dry matter consumption, weight variation, nutrient digestibility, nutrient degradation, passage rate, feed efficiency and economic viability.

2. BIBLIOGRAPHICAL REFERENCES

ALVES, K. S.; CARVALHO, F. F. R.; FERREIRA, M. A. et al. Energy levels in diets for Santa Inês sheep: apparent digestibility. *Revista Brasileira de Zootecnia*, v.32, p.1962-1968, 2003.

ASELTINE, M. S. Corn silage quality can vary depending on hybrid planted. *Feedstuffs*, v.60, p.13-15, 1988.

ASTIGARRAGA, L. Técnicas para la medición del consumo de ruminantes en pastoreo. In: SIMPÒSIO SOBRE AVALIAÇÃO DE PASTAGENS COM ANIMAIS, Maringà: Cooper. Graf. Artes Grâficas Ltda, 1997. Proceedings... p. 1-23.

BAL, M. A.; SHAVER, R. D.; SHINNERS, K. J. et al. Stage of maturity, processing, and hybrid effects on ruminal *in situ* disappearance of whole-plant corn silage. *Animal Feed Science Technoogy*, v.86, p.83-94, 2000.

BERCHIELLI, T. T.; ANDRADE, P.; FURLAN, C. L. et al. Evaluation of internal indicators in digestibility trials. *Revista Brasileira de Zootecnia*, v.29, p.830-833, 2000.

BERCHIELLI, T. T.; OLIVEIRA, S. G.; FEITOSA, W. et al. Estimates of fecal production and total digestibility in cattle using indicators. IN: REUNIÂO ANUAL DA SOCIEDADE BRASILEIRA DE ZOOTECNIA, 42. Goiânia. Anais...: Brazilian Society of Animal Science, 2005.

BURNS, J.C.; POND, K.R.; FISHER, D.S. Measurement of forage intake. In: FAHEY Jr., G.C. (Ed.) *Forage Quality, Evaluation, and Utilization.* p. 494-531, 1994.

CABRAL, L. S.; NEVES, E. M. O.; ZERVOUDAKIS, J. T. et al. Estimates of the nutritional requirements of sheep under Brazilian conditions. *Revista Brasileira de Saùde e Produçao Animal*. v.9, n.3, p.529-542, 2008.

CABRAL, L. S.; VALADARES FILHO, S. C.; DETMANN, E. et al. Consumption and digestibility in cattle fed diets based on tropical forages. *Revista. Brasileira de Zootecnia*, 2006.

COCHRAN, R. C.; ADAMS, D. C.; WALLACE J. D. et al. Predicting digestibility of different diets with internal markers: evaluation of four potential markers. *Journal of Animal Science*, v.63, p.1476-1483, 1986.

COSTA, H. N.; PEREIRA, M. N.; MELO, R. P. et al. Effect of the rumen environment on ruminal in situ degradability of sugarcane. In: WORLD CONFERENCE ON ANIMAL PRODUCTION, 9. Porto Alegre. Proceedings...UFRGS, 2003.

COSTA, M. A. L.; VALADARES FILHO, S. C.; VALADARES, R. F. D. et al. Validation of the NRC (2001) equations for predicting the energy value of feeds under Brazilian conditions. *Revista. Brasileira de Zootecnia,* v.34, p.280-287, 2005.

DETMANN, E.; PAULINO, M. F.; ZERVOUDAKIS, J. T. et al. Chromium and internal indicators in determining the intake of supplemented crossbred steers at pasture. *Revista Brasileira de Zootecnia*, v.30, p.1600-1609, 2001.

DETMANN, E.; SOUZA, A. L.; GARCIA, R.; et al. Evaluation of the "long time" bias of internal indicators in ruminant digestion assays. *Arquivo Brasileiro de Medicina Veterinària e Zootecnia*, v.59, n.1, p.182-188, 2007.

DIAS, M.; DETMANN, E.; LEAO, M. I. et al. Indicators for estimating partial digestibility in cattle. *Revista Brasileira de Zootecnia,* 36: 689-697. 2007.

DIAS, M. et al. Techniques for estimating digestibility and microbial production in cattle. *Revista Brasileira de Zootecnia,* Viçosa, v. 37, n. 6, p. 504-512, Jun, 2008. 1995.

FONDEVILA, M.; CASTRILLO, C.; GASA, J. et al. Rumen-undegradable dry matter and neutral detergent fiber as ratio indicators of digestibility in sheep given cereal straw-based diets. *Journal Agriculture Science*, v.125, p.145- 151, 1995.

PHIPPS, R.H. (Ed.). Improvement of quality traits of maize for grain and silage use. Brussels: Martinus Nijhoff Publ., 1980. p.319-339.

GLINDEMANN, T.; TAS, B. M.; WANG, C. et al. Evaluation of titanium dioxide as an inert marker to estimate fecal excretion in grazing sheep. *Animal Science and Technology Feed* , v.152, p.186-197, 2009.

GREENHALGH, J.F.D. An introduction to herbage intake measurements. In: LEAVER, J.D. Herbage Intake Handbook. Hurley, UK: The British Grassland Society,1982. p.1-10.

HADDAD, C. M.; CASTRO F. G. F. Hay production. In: MANAGEMENT OF TIFTON, COASTCROSS AND ESTRELA PASTURES. Piracicaba: FEALQ, 1998. Proceedings... p.151171.

HEMKEN, R. W.; CLARK, N. A.; GOERING, H. K. et al. Nutritive value of corn silage as influenced by grain content. *Journal Dairy Science,* v.54, p.383-389, 1971.

HILL, G.M.; GATES, R.N.; BURTON, G.W. Forage quality and grazing steer performance from Tifton 85 and Tifton 78 bermuda grass pastures. *Journal of Animal Science,* v.71, p. 3219-3225, 1993.

ÌTAVO, L. C. V.; VALADARES FILHO, S. C.; SILVA, F. F. et al. Intake, ruminal degradability and apparent digestibility of Cynodon grass hays and concentrate feeds using internal indicators. *Revista Brasileira de Zootecnia*, v.31, p.1024-1032, 2002.

LE DU, Y. L. P., PENNING, P. D. Animal based techniques for estimating herbage intake. In: LEAVER,J.D. Herbage intake handbook. Dumfries: The British Grassland Society. p.37-76. 1982.

LIMA, L. P.; VELOSO, C. M.; SILVA, F. F. et al. Cassava bagasse (*Manihot esculenta,* Crantz) in the diet of dairy cows: nutrient intake. *Arq. Bras. Med. Vet. Zootec.,* v.60, p.1004-1010, 2008.

LIPPKE, H.; ELLIS, W. C.; JACOBS, B. F. Recovery of indigestible fiber from feces of sheep and cattle on forage diets. *Journal of Dairy Science*, v.69, n.2, p.403-412, 1986.

LOPES, F. C. F. Consumption of tropical forage by lactating cows under grazing in intensive milk production systems. III SYMPOSIUM ON DAIRY CATTLE NUTRITION AND PRODUCTION: MILK PRODUCTION ON PASTURE. UFMG Veterinary School, Proceedings...2007.

MACÊDO, G. A. R.; VIANA, M. C. M.; OLIVEIRA, J. S. Agronomic and bromatological characteristics of sugarcane varieties in the Alto Paranaiba region, Minas Gerais. In: ANNUAL MEETING OF THE BRAZILIAN SOCIETY OF ZOOTECHNICS, 43, Joao Pessoa. Proceedings... SBZ, 2006. CD-ROM.

MAYES, R.W. The quantification of dietary intake, digestion and metabolism in farm livestock and its relevance to the study of radionuclide uptake. *Science Total Environ.*, v.85, p.29-51, 1989.

MERCHEN, N. R. Digestion, absorption and excretion in ruminants. In: CHURCH, D.C. (Ed.) *The ruminant animaldigestive physiology and nutrition.* Englewood Cliffs: Prentice Hall, 1988. p.172-201.

MICHALET-DOREAU, B.; DOREU, M. Maize genotype and ruminant nutrition. *Science. Alim.,* v.19, p.349-365, 1999.

MINSON, D. J. Forage in ruminant nutrition. San Diego: Academic Press, Inc., 1990. 483p.

MINSON, D.; WILSON, J. R. Prediction of intake as an element of forage quality. In: FAHEY JR., G. C. (Ed.) *Forage Quality, Evaluation and Utilization*. Madison: ASA, CSSA, SSSA, 1994. p. 533-563.

MORAES, K. A. K.; VALADARES FILHO, S. C.; MORAES, E. H. B. K. et al. Consumption of beef heifers fed sugar cane with calcium oxide and different concentrate offerings. In: ANNUAL MEETING OF THE BRAZILIAN SOCIETY OF ZOOTECHNICS, 44, Jaboticabal. Anais... SBZ, 2007. CD-ROM.

MYERS, W. D.; LUDDEN, P. A.; NAYIGIHUGU, V. et al. Technical Note: a procedure for the preparation and quantitative analysis of samples for titanium dioxide. *Journal of Animal Science,* v.82, n.1, p.179-183, 2004.

MYERS, W. D.; LUDDEN, P. A; NAYGIHUGU, V. et al. Excretion patterns of titanium dioxide and chromium oxide in duodenal digesta and feces of sheep. *Pesquisa de Pequenos Ruminantes* , v.63, p.135-141, 2006.

OBA, M.; ALEN, M. S. Effects of brown midrib 3 mutation in corn silage on dry matter intake and productivity of high yielding dairy cows. *Journal Dairy Science.* v. 82, n.1 p.135-142, 1999.

OWENS, F. N.; HANSON, C. F. External and internal markers for appraising site and extent of digestion in ruminants. *Journal of Dairy Science*, v. 75, n. 9, p. 2605-2617, 1992.

PENNING, P. D.; JOHNSON, R. H. The use of internal markers to estimate herbage digestibility and intake. *Journal of Agricultural Science*, v.100, p.127-131, 1983.

PERIPOLLI, V. Use of fecal nitrogen to estimate intake and digestibility in grazing ruminants. Porto Alegre: Faculty of Agronomy, Federal University of Rio Grande do Sul, 2010, 76p. Dissertation (Master's Degree in Animal Science).

PHILIPPEAU, C.; MICHALET-DOREAU, B. Influence of genotype and stage of maturity of maize on rate of ruminal starch degradation. *Animal Feed Science Technology,* v.68, p.25-35, 1997.

PHILIPPEAU, C.; LANDRY, J.; MICHALET-DOUREAU, B. Influence of the biochemical and physical characteristics of the maize grain on ruminal starch degradation. *Journal. Agriculture Food Chemical,* v.46, p.4287-4291, 1998.

PHIPPS, R. H.; WELLER, R. F. The development of plant components and their effects on the composition of fresh and ensiled forage maize. 1. The accumulation of dry matter, chemical composition and nutritive value of fresh maize. *Journal Agriculture Science,* v.92, p.471-483, 1979.

RODRIGUES, A.A.; CRUZ, G.M.; BATISTA, L.A.R. et al. Quality of ten varieties of sugar cane as feed for cattle. In: REUNIAO ANUAL DA SOCIEDADE BRASILEIRA DE ZOOTECNIA, 42, Goiânia: SBZ; FEALQ, 2001. Anais... p.1111-1113.

RODRIGUES, A.A.; CRUZ, G.M.; BATISTA, L.A.R. et al. Quality of nine varieties of sugar cane as feed for cattle. In: REUNIAO ANUAL DA SOCIEDADE BRASILEIRA DE ZOOTECNIA, 43. Joao Pessoa: SBZ; Anais... 2006. CD-ROM.

RODRIGUEZ, N. M.; GUIMARÂES JUNIOR, R. Utilization of agro-industry by-products in the feeding of dairy cows. In: SIMPÒSIO MINEIRO DE NUTRIÇÂO DE GADO DE LEITE, 3, Belo Horizonte: EV/UFMG, Anais... 2005. p.65- 91.

RODRIGUEZ, N. M; SALIBA, E. O. S.; GUIMARÂES JÛNIOR, R. Use of indicators to estimate pasture consumption and digestibility. IN: 43ª REUNIÂO ANUALO DA SOCIEDADE BRASILEIRA DE ZOOTECNIA, Joao Pessoa: SBZ; Anais... 2006. CDROM.

SALIBA, E. O. S. Chemical and microscopic characterization of lignins from corn and soybean residues exposed to rumen degradation and their effect on the digestibility of structural carbohydrates. Belo Horizonte: Veterinary School of the Federal University of Minas Gerais, 1998, 251p. Thesis (Doctorate in Animal Science).

SALIBA, E. O. S.; RODRIGUEZ, N. M.; PILO-VELOSO, D. Purified lignin extruded from Eucalyptus grandis (PELI), used as an external marker in digestibility trials in various animal species. In: WORLD CONFERENCE ON ANIMAL PRODUCTION, 9, Porto Alegre: WAAP/ALPA/SBZ/UFRGS; Anais... 2003.

TITGEMEYER, E. C. Design and interpretation of nutrient digestion studies. *Journal Animal Science*, v. 75, p. 2235-2247, 1997.

TITGEMEYER, E. C.; AMENDARIZ, C. K.; BINDEL, D. J. et al. Evaluation of titanium dioxide as a digestibility marker in cattle. *Journal Animal Science* 79:1059-1063. 2001.

VALADARES FILHO, S. C. Nutrition, feed evaluation and feed composition tables for cattle. In: Reuniao Anual da Sociedade Brasileira de Zootecnia, Viçosa: SBZ; Anais... p.267-337. 2000.

VALADARES FILHO, S. C.; MAGALHAES, K. A.; ROCHA Jr.; V. R. et al. Tabelas brasileiras de composiçao de alimentos para bovinos. 2.ed. Viçosa: DZO/UFV, 2006. 329p.

VAN SOEST, P. J. Nutritional ecology of the ruminant. New York: Cornell University Press, 1994, 476p.

ZEOULA, L. M.; PRADO, I. N.; DIAN, P. H. M. et al. Fecal recovery of internal markers evaluated in ruminants. *Revista Brasileira de Zootecnia*, v.31, n.4, p.1865- 874. 2002.

CHAPTER 3 - USE OF DIFFERENT INDICATORS IN DETERMINING CONSUMPTION, DIGESTIBILITY AND FECAL PRODUCTION IN BOVINS, COMPARED WITH TOTAL FECAL COLLECTION

1. INTRODUCTION

The search for new techniques to determine the nutritional value of food has been the aim of several studies involving animals. In this context, the study of digestibility, consumption and fecal production is essential when formulating animal diets. Therefore, they are limiting factors in the animal's productive efficiency, depending on the characteristics of the food, the environment and its interactions, and the inherent factors of the animal itself (Peripolli., 2010).

Direct determination of fecal production, known as *in vivo* total fecal collection, is an expensive process and requires strict control of the animal's fecal consumption and production. Although *in vivo* procedures have their limitations in some cases, they remain the benchmark for both food evaluation and the validation of alternative determination methods.

In order to develop alternatives that make it possible to estimate fecal production, digestibility, and indirectly consumption, indicator techniques have been proposed, which are chemical monitors according to Owens and Hanson (1992) for quantitative and qualitative determination of physiological and nutritional phenomena.

Indicators can be classified as internal, represented by indigestible substances naturally present in some component of the diet, and external, when added to the diet or supplied orally.

2. OBJECTIVES

The objectives of this work were to determine the consumption, digestibility and fecal production of Holstein-Gyr mixed-breed heifers using different techniques. For this purpose, total collection (standard methodology), the external indicators LIPE®, titanium dioxide and the internal indicators indigestible dry matter and indigestible neutral detergent fiber at 264 hours of *in situ* incubation, and indigestible dry matter and indigestible neutral detergent fiber at 72 and 144 hours of *in vitro* incubation were used.

3. MATERIAL AND METHODS

3.1 EXPERIMENTAL LOCATION, ANIMALS AND FACILITIES

The experiment was conducted on the experimental farm of the Central de Ensino e

Desenvolvimento Agràrio de Florestal (CEDAF), in the city of Florestal (19° 53' 20"S 44° 25' 58"W), in Minas Gerais, 60 km from the capital, Belo Horizonte. The climate in this region is considered to be high altitude tropical. The laboratory analyses were carried out at the Animal Nutrition Laboratory of the Zootechnics Department of the Federal University of Minas Gerais (LABNUTRI/EV-UFMG). Nine Holstein x Gir mixed-breed heifers were used, fistulated in the rumen, with an average initial live weight of 480 kg and 24 months of age. They were kept in individual *tie stalls with* concrete floors, covered, equipped with individual automatic drinking fountains and troughs for feeding the diet. Figure 1 partially illustrates the facilities.

Figurei. Experimental facilities.

Source: Personal archive

The experiment lasted 45 days, divided into three 15-day periods. The animals were weighed, identified and dewormed at the start of the experiment and kept for 10 days to adapt to the experimental diet and facilities and five days for collection (feces, food offered and leftovers). Weights were taken at the end of each period in order to adjust the diets.

3.2 STATISTICAL EXPERIMENTAL DELINEATION

The statistical experimental design used was a simultaneous triple Latin square with subdivided plots. The data was evaluated using the SISVAR program, using the SNK test at 95% probability. Table 1 describes the analysis of variance.

Table 1: Analysis of variance

SOURCES OF VARIATION	DEGREES OF FREEDOM
Total	26
Latin Square	2
Period	2
Animal	6
Diet	2

Error A	14
SOURCE OF VARIATION	
Total	242
Indicators	8
Interaction (diet x indicators)	16
Error B	192

Where: Yijklm = μ + Pi + Sj + Dk + Zl + Wm + AIij + eij;

μ= general average; Pi= period effect; SJ= animal effect; Dk= diet effect; Zl= sub-plot effect; Wm= indicator effect; PSijklm = interaction between diet and indicator effects; Eij = random error.

3.3 EXPERIMENTAL DIETS

Three treatments were evaluated, consisting of different bulk sources in the experimental diets: corn silage (*Zea mays*, L.), Tifton grass hay (*Cynodon spp.*) and sugar cane (*Saccharum spp.*). The diets were supplemented with concentrate based on cornmeal, soybean meal and mineral premix in order to meet the nutritional requirements of the animals in this study and were balanced to be isonitrogenous (14% CP), consisting of bulk and concentrate in a 75:25 ratio.

The dietary ingredients were weighed and mixed manually twice a day to be fed at 8:00 am and 4:00 pm as a complete diet, with consumption measured daily, allowing for 5 to 10% of daily leftovers. The ingredient and chemical-bromatological composition of the experimental diets is described in Tables 2 and 3.

Table 2. Composition of the experimental diets, in ingredients, based on % of Dry Matter (DM).

Ingredients	Diets		
	Corn silage	Sugar cane	Tifton hay
Corn silage	74,1	-	-
Sugar cane	-	73,8	-
Tifton hay	-	-	74,5
Ground corn	16,3	-	21,5
Soybean meal	8,9	25,5	3,3

Minerals and Vitamins	0,7	0,7	0,7

[1]Minerals and Vitamins: 150 g Ca; 90 g P; 17.78 g Mg; 15.4 g S; 114 g Na; 1,500 mg Mn; 4,000 mg Zn; 1,200 mg Cu; 145 mg Co; 90 mg I; 38 mg Se; 900 mg F; 200,000 IU Vit A; 50,000 IU Vit D and 1500 mg Vit E per kg of mixture.

Table 3. Composition of experimental diets, in nutrients, based on % of Dry Matter.

Nutrient	Diets		
	Corn silage	Sugar cane	Tifton hay
Crude protein	13,9	13,5	13,8
Neutral detergent fiber	51,3	50,8	61,8
Ashes	5,2	5,4	6,4
Ethereal extract	2,7	1,1	2,3
Non-fibrous carbohydrates	26,9	29,2	15,7
Total carbohydrates	78,2	80,0	77,5
Dry matter (%MN)	31,5	37,1	89,4

3.4 REAL CONSUMPTION

The diet provided and each animal's leftovers were recorded daily before morning feeding. A composite sample was formed for each heifer in each period, based on the dry matter of the daily feed collections and leftovers. Daily dry matter intake (DMI) was calculated by multiplying the daily natural matter intake (DMI) of each feed, between the 10th and 15th day of each period, by its respective dry matter content, followed by subtraction of the dry matter leftovers, according to the formula described below:

CMS= (CDMN x % DM) - DM (leftovers)

The daily nutrient intake per animal was calculated by multiplying the dry matter offered for each ingredient by its respective nutrient content. From the total nutrients offered, each animal's daily surplus of the same nutrient was subtracted.

During the collection period, samples of each animal's roughage, concentrates and leftovers were collected daily, packed in plastic bags and stored in a freezer at -20°C. At the end of the period, pooled samples were taken from each animal.

3.5 Apparent and *in vitro* digestibility

The apparent digestibility of the feed was determined by collecting all the feces and controlling the feed offered and the animals' leftovers over a period of five consecutive days, between the 10th and 15th day of each period. Every 24 hours, the faecal production obtained from the spontaneous defecation of each animal was weighed, homogenized and samples of approximately 200g of faeces were taken. During the total collection period, the feces were continuously frozen and a composite sample was formed per heifer based on dry matter, per period. The average fecal production was obtained from the average of the five collection days.

In addition to the apparent digestibility, the *in vitro* dry matter digestibility of the different diets during the different periods was also carried out. Samples of the diets offered were collected daily and individually (per animal), and at the end of each period, composite samples were taken.

The *in vitro* digestibility of dry matter was carried out according to Tilley and Terry (1963). The results obtained through *in vitro* digestibility are shown in Table 4.

Table 4. *In vitro* Digestibility of Dry Matter expressed in %, of the diets based on Corn Silage (SM), Sugar Cane (CA) and Tifton Hay (FT) in the three experimental periods.

Periods	Diets		
	SM	CA	FT
1	68,10	67,01	66,99
2	69,8	69,01	70,00
3	69,6	69,01	69,99

3.6 USE OF INDICATORS TO DETERMINE CONSUMPTION, FECAL PRODUCTION AND DIGESTIBILITY

To estimate consumption, fecal production and digestibility, the internal indicators used were indigestible dry matter (MSi), indigestible neutral detergent fiber (FDNi), and the external indicators LIPE® and titanium dioxide (TiO).2

3.6.1SUPPLY OF EXTERNAL INDICATORS LIPE® AND TIO2

The external indicator LIPE® was supplied in the form of capsules, at a dose of 0.5g per heifer per day, via the rumen cannula, with two days for adaptation plus supply throughout the collection period (five days), in total seven capsules/animal/period were supplied.

TiO_2 was fed to the animals in cellulose capsules containing 10g, directly through the rumen cannula, once a day for seven days of adaptation and four days of collection, totaling 11 capsules. Table 5 shows the period of supply of the indicators during each experimental stage.

Table 5. Administration of external indicators LIPE® and TiO_2 by day and collection period.

Supply of external indicators															
Days	1	2	3	4	5	6	7	8	9	10	11	12	13	14	15
LIPE®									x	x	xc	xc	xc	xc	c
TiO2				x	x	x	x	x	x	x	xc	xc	xc	xc	c

*The tables marked with the letter x correspond to the period in which the indicators were provided, and the tables marked with the letters "xc" or just the letter "c" correspond to the collection period.

At the end of the 15th day, a new period began, with new treatments, but the period of adaptation of the indicators was maintained.

3.6.2OBTAINING THE INTERNAL INDICATORS, MSi AND FDNi

Samples of all the roughage, concentrates, leftovers and feces were incubated in an animal's rumen for 264 hours as suggested by Casali et al. (2008), to obtain the indigestible fractions of the feed MSi and FDNi, *in situ, in* addition to the *in situ* incubations, *in vitro* incubations were carried out at 72 hours as suggested by (Berchielli et al., 2000) and 144 hours (Conchran et al., 1986; Berchielli et al., 2000) for the same internal indicators.

To carry out the rumen incubation (*in situ*), non-woven fabric (NW) bags were used, with a density of $100g/cm^2$. The samples of feed, leftovers and feces were ground in a Willey mill with a 1 mm sieve, and placed in the TNT filter bags, in a quantity of 20 to 25 mg of dry matter per cm^2 of surface, approximately 1 g of sample, respecting both sides of the containers. Figures 2, 3 and 4 illustrate the basic procedures for the rumen incubation process. After 264 hours of *in situ* incubation, the filter containers were removed from the rumen and rinsed with running water. They were then taken to the oven and, once dry, weighed and the residue used to determine the DMi. The FDNi was determined sequentially to the MSi, where the containers were washed with neutral detergent solution, then dried in an oven, and finally weighed, and the residue was used to determine the indigestible fraction of the FDN.

Figure 2: Filter containers (TNT).

Source: Personal archive

Figure 3: Rumen incubation process

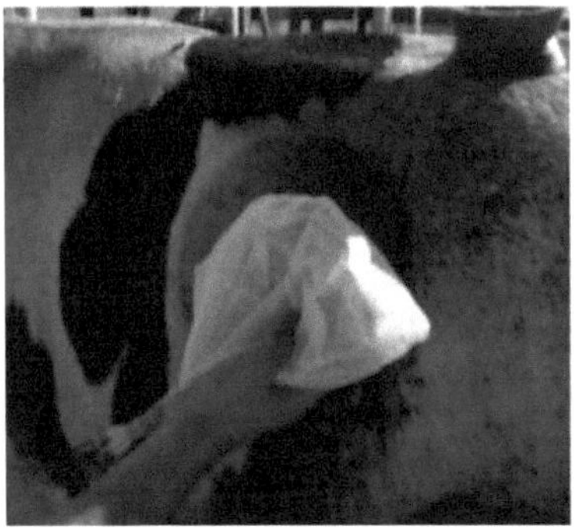

Source: Personal archive

Figure 4: Removal of the incubated material.

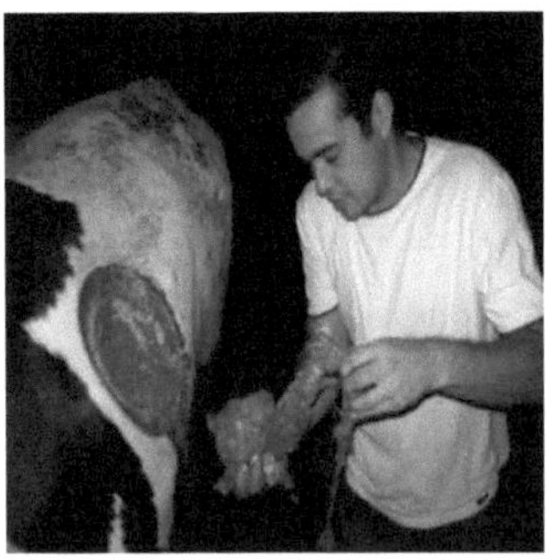

Source: Personal archive

For the *in vitro* determination of MSi at the different times, 72 and 144 hours, the samples were also ground in a Willey mill with a 1 mm sieve, but placed in standardized ANKON filter bags (F-57), respecting 0.5g of sample per bag.

The rumen fluid used for *in vitro* determination of the indigestible fractions was taken from a male mestizo bovine, fistulated in the rumen, housed at the UFMG Veterinary School, receiving a balanced diet of roughage and concentrate. The samples of NDF determined by the *in vitro* methodology, at 72 and 144 hours, were taken sequentially after obtaining the MSi, where the filter containers were washed with neutral detergent solution, and after drying, they were weighed and their residue was used to determine the NDFi.

3.7 CHEMICAL ANALYSIS

The food samples, leftovers and feces were pre-dried at 55° C. They were then ground in a Willey mill with a 1 mm sieve and stored in plastic containers for chemical analysis. The dry matter (DM), mineral matter (MM), crude protein (CP) and ethereal extract (EE) contents were analyzed according to INCT (2012) recommendations. The contents of neutral detergent fiber (NDF) and acid detergent fiber (FDA) were determined according to the techniques described by Silva (1998).

The percentages of non-fibrous carbohydrates (NFC) and total carbohydrates (TC) were obtained using the equations proposed by Sniffen et al. (1992): NFC = 100 - (%FDN + %PB + %EE + %Ashes) and TC =100 - (%PB + %EE + %Ashes). Total carbohydrates (TC) and total digestible nutrients (NDT) were calculated according to the method described by Sniffen et al. (1992), where: TC = 100 - (%PB + %EE + % Ash). The NDT of the diets was calculated according to the equation described by the NRC (2001), in which NDT = %digestible BNP + %digestible FNP + %digestible NCF + 2.25*%digestible EE.

The LIPE® indicator was analyzed using infrared spectroscopy on Varian-800® equipment with Fourier transform (FT-IV) (Saliba, 2005).

To determine titanium dioxide, procedures were carried out according to Myers et al. (2004) using molecular electronic spectroscopy, based on colorimetry.

To obtain fecal production (FP) from the external indicators, the following formulas were used:

$$PF = \frac{\text{LIPE® supplied (g)}}{(\text{LIPE® in feces x fecal DM})}$$

or

$$PF = \frac{TiO_2 \text{ supplied (g)}}{(TiO_2 \text{ in feces x fecal DM})}$$

To obtain consumption from the external indicators LIPE® and TiO_2 , the following formula was used:

$$\text{Consumption} = \frac{PF}{1 - \text{Digestibility}}$$

The following formula was used to obtain consumption from internal indicators:

$$\text{Dry matter intake (g)} = \frac{\text{Excretion of internal indicator in faeces (g)}}{\text{\% Internal Indicator in Ingested Food}}$$

The following formula was used to obtain digestibility from the internal indicators:

$$\text{Digestibility} = 100 - 100 \times \left[\frac{\text{\% Indicator in food}}{\text{\% Indicator in feces}} \right]$$

4. RESULTS AND DISCUSSION

4.1 TRUE INTAKES OF THE DIFFERENT NUTRIENTS IN THE DIETS

Average daily intakes of dry matter (DM), organic matter (OM), crude protein (CP), ether extract (EE), neutral detergent fiber (NDF), non-fibrous carbohydrates (NFC), total carbohydrates (CHO) and total digestible nutrients (TND), for the total diet. The consumption of dry matter (DM), neutral detergent fiber (NDF) and indigestible neutral detergent fiber (INDF) from the roughage in kg per day and the consumption of DM and NDF as a function of live weight are shown in Table 6.

Table 6. Nutrient intake in heifers receiving diets based on corn silage (SM), sugar cane (CA) and Tifton hay (FT).

Variables	Diets			EPM[1]	p2
	SM	CA	FT		
	Total Diet Consumption (kg/day)				
Dry matter	10,6	10,2	9,56	0,52	0,33
Organic matter	10,2	9,7	8,9	0,50	0,25
Crude protein	1,4	1,6	1,4	0,07	0,07
Ether extract	$0{,}3^{a}$	$0{,}1^{c}$	$0{,}2^{b}$	0,01	<0,01
Neutral detergent fiber	5,2	4,8	5,7	0,28	0,16
Non-fibrous carbohydrates	$3{,}2^{a}$	$3{,}1^{a}$	$1{,}7^{b}$	0,17	<0,01
Total digestible nutrients	7,6	7,0	6,5	0,42	0,26
	Feed intake (kg/day)				
Dry matter	8,0	7,6	7,1	0,39	0,31
Neutral detergent fiber	3,9	3,6	4,3	0,22	0,14
		Consumption (%			

		BW)			
Dry matter	2,2	2,1	2,0	0,12	0,44
Neutral detergent fiber	1,1	1,0	1,2	0,07	0,26

EPM[1] = Standard error of the mean/ *P* = Probability value for the diet effect, Means followed by different letters, lower case in the row and upper case in the column, represent differences according to the SNK test ($P<0.05$).

Dry matter intake (kg/day) did not differ between the experimental diets (Table 6). Despite the absence of any significant difference, the animals given diets containing corn silage ingested an average of 0.5 kg of organic matter compared to those given sugarcane and 1.2 kg compared to those given hay as a roughage, a fact that can be attributed to the higher consumption of NDF from the roughage in the diet with tifton hay, in agreement with the report by Cabral et al. (2006), despite the small trend ($P=0.14$), as confirmed by Cavalcante et al. (2004). When corn silage and sugarcane were compared, no significant differences were found, a fact also verified by Costa et al. (2005), Rangel et al. (2010).

When consumption was determined as a function of live weight, a general average dry matter consumption of 2.1% was recorded, which is close to the value recommended by the NRC (2001) for this category of animal, in their physiological and weight conditions. As for the consumption of NDF as a function of live weight, this was statistically the same between the diets studied. The results found in this study for NDF intake as a function of BW, regardless of the roughage used, are within this limit, with the exception that for diets based on sugarcane or silage the values were lower than those described for maximum DM intake, a fact also observed by Mendonça et al. (2004) and Costa et al. (2005) when they tested diets containing corn silage or sugarcane as the exclusive roughage. For the hay-based diet, this percentage was exactly the same as the 1.2% limit proposed by the aforementioned researcher. According to Van Soest (1994), if the cell wall limits intake, it has a constant maximum intake. However, in this study, the level of NDF in the diet was not a limiting factor in controlling voluntary intake. As the V:C bulk:concentrate ratio (75:25) was the same for all the diets, it can be inferred that in all the diets intake was controlled by the animal's energy demand, in line with the considerations made by Conrad et al. (1964) and Van Soest (1994), who reported that in diets with digestibility above 66%, as observed in this study for all the diets, dry matter intake is controlled by physiological factors.

Sugarcane has been negatively correlated with dry matter intake, not only because of the indigestible fiber fraction, but also because of the low digestion rate of potentially degradable fiber, which has a high rumen repletion effect (Pereira et al., 2000). According to Van Soest (1994), the consumption of NDF is what best represents the consumption capacity for diets based on roughage and other feeds with medium and low amounts of concentrate, as it is the constituents of the cell wall that occupy the most space in the rumen.

The consumption of organic matter, crude protein, total carbohydrates and total digestible nutrients followed the same trend as the consumption of dry matter, and did not differ between the diets (Table 6), a fact that was also evidenced in the studies by Cavalcante et al. (2004), where they evaluated the replacement of hay with silage for steers

Dias (2009), when evaluating diets with sugarcane or corn silage for heifers in diets with a V:C ratio (100:0 or 75:25), also found no differences in the consumption of OM and CP.

Average intakes of NFC were higher in the corn silage and sugarcane diets, which is explained by the higher content of this nutrient in these diets (Table 6). The same was observed for the EE fraction, whose highest consumption was recorded for the silage diet, followed by hay and finally the sugarcane diet, results which reflect the percentage of this fraction in the respective diets. A similar fact was observed by Cabral et al. (2006) when evaluating different feedstuffs.

An extensive discussion of the separate nutrient intakes was given above. We thought this discussion was interesting since the indigestible fibrous fraction will be used as one of the tools for determining the digestibility of diets.

4.2 FECAL PRODUCTION OBTAINED BY TOTAL COLLECTION AND DETERMINED WITH THE USE OF INDICATORS

Apparent fecal production (AFP) through total fecal collection and estimates of fecal production determined by the indicators LIPE® , titanium dioxide, MSi and FDNi *in situ* incubated for 264 hours (h), MSi and FDNi *in vitro* incubated for 72 hours and 144 hours (h) are shown in Table 7.

Table 7 - True fecal production and estimated by different indicators, expressed in kg of dry matter in different diets, corn silage (SM), tifton hay (FT), sugar cane (CA).

Indicator	Diet			Average
	SM	FT	CA	
Total collection	3,17	2,89	3,07	3.05 AB
LIPE®	3,15	3,07	2,87	3.03 AB
TiO_2	3,08	2,97	2,95	2.99 AB
MSi 264 (h)	2,95	2,95	3,00	2.97 AB
FDNi 264 (h)	2,40	2,72	2,19	2,44 C
MSi 72 (h)	3,16	2,72	3,54	3.14 AB

FDNi72 (h)	3,33	3,28	3,42	3,35 A
MSi144 (h)	3,05	2,92	3,03	2.99 AB
FDNi 144 (h)	2,94	2,36	3,05	2,78 B

Averages followed by different letters, lower case in the row and upper case in the column, represent differences using the SNK test (P<0.05). HP (%) = 20.56.

Analyzing the data in Table 7, it can be seen that there was no interaction between the indicators and the different diets, corn silage, tifton hay and sugar cane. This fact disagrees with that found by Berchielli et al. (2005) where they reported a different behavior of the indicators according to each of the bulks studied, indicating that, possibly, the reason for this difference is associated with the constitution of the fiber and its influence on the rate and extent of degradation of its own components.

As there was no interaction between indicator and diet, the discussion is based on the marginal averages imposed by all the indicators, including the total collection, which is the basic parameter for comparisons.

The fecal production estimated by LIPE®, when compared to the total collection, shows no significant average difference by the SNK test with 95% probability. In other words, the data presented for the indicator showed consistency in determining fecal production when compared to the total collection.

The fecal production data found using the LIPE indicator® was 3.03 kg while the total collection was 3.05, thus corroborating the data obtained by Saliba et al. (2003), working with ruminants, compared total collection of feces and estimates with LIPE®, in a trial evaluating Tifton 85 hay in sheep diets, where they verified the results obtained using the two forms of measurement (indicator versus total collection) which resulted in similar responses for fecal production values of 365.39 g/day and 383.07 g/day, respectively.

Paixao et al (2007), evaluating the variation in fecal excretion of various indicators, including LIPE®, compared to total collection in steers, concluded that this indicator can satisfactorily estimate fecal production, promoting fecal recovery close to 100%.

Silva et al. (2008), Merlo et al. (2008) and Silva et al. (2008b), working with sheep, observed no difference between the data found by the total feces collection method and LIPE® in determining the digestibility of diets with increasing levels of cashew nuts, coconut bran and babassu cake, respectively.

The fecal output estimated by titanium dioxide was 2.99 kg, while the true fecal output was 3.05 kg.

These data show that there was no significant difference in the SNK test with a 95% probability.

These results agree in part with the data obtained by Titgemeyer et al. (2001), when they conducted three experiments with steers, evaluating the accuracy of the Cr2O3 and TiO2 indicators. In the first experiment, the animals were fed diets based on hay and concentrate and in the others on diets based on dry rolled corn and soybean meal. Only in the first experiment did TiO2 estimate fecal excretion values similar to those obtained from total fecal collection, and in the others, the values were underestimated.

Glidemann et al. (2009), evaluating fecal production estimated by titanium dioxide in sheep, found fecal production and fecal recovery values similar to those obtained with total fecal collection.

Marcondes et al. (2006a) used mixed-breed heifers fed sugar cane and concentrate to evaluate TiO_2 in estimating fecal production obtained in three or six days of fecal collection. These researchers concluded that TiO_2 was sufficiently capable of estimating total collection after only three days when compared to actual fecal production.

Ferreira et al. (2009a) carried out an experiment under similar conditions and also concluded that only three days of collection for titanium dioxide would be sufficient to determine fecal production.

The indigestible fractions of the feed were also used to determine fecal production and are shown in Table 7. Fecal production estimated by MSi *in* 264 hours of *in situ* incubation was similar to true fecal production, with values of 2.97 kg and 3.05 kg respectively. These data corroborate the time proposed for obtaining the indigestible fractions of feed according to (Casali et al., 2008; Sampaio et al., 2011).

In this experiment, *in situ* incubations were carried out for 264 hours, which is not recommended by INCT-CA (2009), which is 240 hours, in order to evaluate dietary fibers over a longer period of time. It was found that this longer time was capable of estimating fecal production from MSi, similar to the total collection in the three diets.

Zeoula et al. (2009), when comparing different feeds, incubation times and animal species (cattle and buffalo), found an interaction between species and feed. In the same study, they recommended incubation times of at least 214 hours for buffalo and 224 hours for cattle to obtain indigestible dry matter (IDM).

The data found for fecal production obtained with the FDNi indicator, *in* 264 hours of *in situ* incubation for the SNK test, with a probability of 95%, showed a significant difference ($P<0.05$) when compared to the total collection, where the values found of 2.44 kg for FDNi 264 hours incubated underestimated the actual fecal production values, where the values were 3.05 kg. Some factors such as particle size and sequential analysis may have influenced the underestimated values for NDF. The samples were ground to 1mm, so successive treatments after rumen incubation, i.e. filtering with neutral detergent, may have had some influence on the estimate of fecal production.

In vitro tests were carried out to obtain the indigestible fractions of the food. Incubation times were 72 hours and 144 hours, as proposed by Conchran et al., 1986; Berchielli et al., 2000; Freitas et al., 2002; Oliveira Júnior et al., 2004; Cabral et al., 2008.

Fecal production estimated by MSi at 72 and 144 hours of *in vitro* incubation did not differ statistically from total fecal collection by the SNK test at 95% probability. These results partly agree with the data obtained by Berchielli et al. (2000), who carried out trials with incubation times of 72 and 144 hours, and observed that the shortest incubation time did not reproduce the indigestible fraction of the indicator, highlighting the need for longer incubation times.

The fecal production estimated by the FDNi indicators at 72 and 144 hours of *in vitro* incubation behaved differently when compared to the total collection. The NDF incubated for 72 hours showed no significant difference compared to the data obtained from the total collection, with a fecal production of 3.35 kg compared to 3.05 kg obtained from the total collection.

These data go against the data obtained by Berchielli et al. (2000), where 72 and 144 hours of incubation were used; in this study, 72 hours was not enough to estimate faecal production through NDFi.

When incubated *in vitro* for 144 hours, the values underestimated fecal production compared to the total collection, the values found were 2.78 kg and 3.05 kg respectively. However, as this is an *in vitro* incubation, special attention should be paid to the duration of the inoculum's activity, given the accumulation of fermentation end products and the decrease in the pH of the medium (Traxler, 1997).

4.3 ACTUAL AND ESTIMATED CONSUMPTION BY INDICATORS

The actual consumption (CR) and the estimated consumption determined by the indicators LIPE®, titanium dioxide, MSi and FDNi *in situ* incubated for 264 hours (h), MSi and FDNi *in vitro* incubated for 72 hours and 144 hours (h) are shown in table 8.

Table 8. Actual consumption and estimated consumption by indicators expressed in kg of dry matter in different diets, corn silage (SM), tifton hay (FT), sugar cane (CA).

Indicator	Diet			Average
	SM	FT	CA	
Actual consumption	10,6	9,56	10,2	10.16 AB

LIPE®	10,24	9,79	9,43	9.82 AB
TiO_2	9,98	9,47	9,55	9.66 AB
MSi 264 (h)	10,28	9,47	9,66	9.80 AB
FDNi 264 (h)	10,49	10,11	10,07	10,23 A
MSi 72 (h)	9,32	8,91	8,05	8,77 C
FDNi72 (h)	9,39	9,09	9,39	9.29 BC
MSi144 (h)	9,89	9,50	9,43	9.61 AB
FDNi 144 (h)	10,18	9,79	10,39	10.12 AB

Averages followed by different letters, lower case in the row and upper case in the column, represent differences using the SNK test ($P<0.05$). CV (%) = 10.17

Interpreting Table 8, *in* relation to consumption, only the MSi indicator at 72 hours of *in vitro* incubation differed statistically ($P<0.05$) from true consumption.

The consumption estimated by the other indicators, LIPE®, TiO_2, MSi and FDNi 264 hours incubated *in situ*, MSi 144 hours and FDNi 72 and 144 hours incubated *in vitro,* were statistically similar to the real consumption, evaluated by the SNK test with 95% probability

Oliveira et al. (2005) reported that LIPE®, administered in a single dose (15:00 h) in capsules containing 0.5 g, proved to be an external indicator capable of satisfactorily estimating the DM intake of Nelore steers managed under grazing conditions on Brachiaria brizantha cv. Marandù.

In a study carried out by Silva et al. (2010) LIPE® was effective in predicting the consumption of heifers fed sugar cane and urea. Silva et al. (2011) demonstrated the efficiency of LIPE® in a study with lactating cows, in which no significant difference was found between the dry matter intake estimated by the indicator and the actual intake estimated by total collection.

Lima et al, (2008) in a trial testing chromic oxide and LIPE® in estimating dry matter intake by beef cattle, recommended LIPE® as the most reliable option for indirectly determining dry matter intake in grazing animals.

Pina (2008) also found similarities between concentrate consumption and digestibility, using the indicators chromic oxide and titanium dioxide, working with Nelore heifers fed hydrolyzed sugar cane and different levels of lime when compared to the data obtained by true consumption.

The data found for MSi and FDNi at 264 hours incubated in the rumen, corroborate the data found by Casali et al., 2008; Sampaio et al., 2011) who state that 264 hours is the ideal time to obtain the indigestible fractions.

The MSi data incubated *in vitro* for 72 hours produced results below the data obtained by measuring actual consumption, i.e. they underestimated the data. In view of these results, the work carried out by Berchielli *et al.* (2000) corroborates the data found, where 72 hours of *in vitro* incubation was not enough to represent the value of real consumption.

Torres *et* al. (2009), working with growing dairy cattle fed forage palm, recommended a rumen incubation period of 144 hours for MSi.

For *in vitro* incubations, the 144-hour time represented the consumption estimate from the internal indicators MSi and FDNi well, showing 9.61 and 10.12 kg of dry matter respectively, compared to the actual consumption of 10.14 kg of dry matter.

These data corroborate Conchran et al., 1986; Berchielli et al., 2000; Freitas et al., 2002; Oliveira Júnior et al., 2004; Cabral et al., 2008, who state that 144 hours of *in vitro* incubation is the ideal time for obtaining internal indicators.

4.4 APPARENT DIGESTIBILITY AND ESTIMATED BY DIFFERENT INDICATORS

The apparent digestibility (AD) and digestibility determined by the indicators LIPE®, titanium dioxide, MSi and FDNi *in situ* incubated for 264 hours (h), MSi and FDNi *in vitro* incubated for 72 hours and 144 hours (h) are shown in table 9.

Table 9. Apparent digestibility (A.D.) and digestibility estimated by different indicators. estimated by indicators expressed as a percentage in relation to different diets, corn silage (SM), tifton hay (FT), sugar cane (CA).

Indicator	Diet			Average
	SM	FT	CA	
DA	70,17	69,36	69,84	69,79 A
LIPE®	69,53	67,74	71,22	69,49 A
TiO_2	70,42	68,66	70,90	69,99 A
MSi 264 (h)	70,91	68,52	68,86	69,43 A
FDNi 264 (h)	73,91	72,79	72,35	73,01 A
MSI 72 (h)	54,62	61,05	61,22	58,96 C
FDNI72 (h)	64,23	63,93	63,45	63,88 B

MSI144 (h)	68,82	69,21	67,84	68,62 A
FDNI 144 (h)	71,01	72,32	70,27	71,20 A

Averages followed by different letters, lower case in the row and upper case in the column, represent differences using the SNK test ($P<0.05$). CV (%) = 7.07

Interpreting the data shown in Table 9, it is clear that only the DMi and NDFi estimated *in* 72 hours of *in vitro* incubation were not enough to determine digestibility *in* comparison with actual digestibility. These data corroborate Conchran et al., 1986; Berchielli et al., who state that 72 hours is insufficient time to determine the indigestible fractions of feed.

Saliba et al. (2003) found that there was no difference in the DM digestibility of Tifton grass hay when they compared the data obtained through total collection and LIPE® (63.23 vs 64.78%, respectively).

Ferreira et al. (2009) evaluated the internal indicators indigestible neutral detergent fiber (INDF) and acid detergent fiber (IADF), obtained by incubation *in situ* for 144 hours, and the external indicators chromic oxide (Cr 2O3), titanium dioxide (TiO2) and LIPE® in two total feces collection schemes (3 or 5 days) for estimating digestibility in cattle. They concluded that nutrient digestibility, when estimated using FDAi and external indicators (chromic oxide, titanium dioxide and LIPE® , did not differ significantly from that determined by total collection of feces. With regard to NDFi, the results were not consistent.

5. FINAL CONSIDERATIONS

The external indicators LIPE® and titanium dioxide (TiO2) proved capable of estimating consumption, digestibility and fecal production when compared to the baseline methodology, total fecal collection. With regard to the internal indicators MSi and FDNi incubated *in situ* for 264 hours, the general results of their evaluations allow us to conclude that they are also efficient in determining digestibility, consumption and fecal production. These same indicators, when obtained through *in vitro* tests, by

72 hours of incubation did not show any potential for the estimates in question when compared to total collection. However, when incubated for a period of 144 hours, *in vitro,* the results are statistically similar to the real data in estimating consumption, digestibility and fecal production.

The diets tested had no influence on the fecal recovery of the external indicators or on obtaining the indigestible fractions of the food, demonstrating that the results obtained from the indicators studied are independent of the diet to which the animal is adapted.

6. BIBLIOGRAPHICAL REFERENCES

ANDRADE, M. A. F.; PEREIRA, M. N. Performance of hostein heifers on fresh sugarcane as the only dietary forage. *Journal of Dairy Science*, v.82, p.91, 1999.

BERCHIELLI, T. T.; ANDRADE, P.; FURLAN, C. L. et al. Evaluation of internal indicators in digestibility trials. *Revista Brasileira de Zootecnia*, v.29, p.830-833, 2000.

BERCHIELLI, T. T.; OLIVEIRA, S. G.; FEITOSA, W. et al. Estimates of fecal production and total digestibility in cattle using indicators. IN: REUNIÂO ANUAL DA SOCIEDADE BRASILEIRA DE ZOOTECNIA, 42. Goiânia. Anais...: Brazilian Society of Animal Science, 2005.

CABRAL, L. S.; NEVES, E. M. O.; ZERVOUDAKIS, J. T. et al. Estimates of the nutritional requirements of sheep under Brazilian conditions. *Revista Brasileira de Saùde e Produçao Animal.* v.9, n.3, p.529-542, 2008.

CABRAL, L. S.; VALADARES FILHO, S. C.; DETMANN, E. et al. Consumption and digestibility in cattle fed diets based on tropical forages. *Revista. Brasileira de Zootecnia*, 2006.

CASALI, A. O.; DETMANN, E.; VALADARES FILHO, S. C. et al. Influence of incubation time and particle size on the levels of indigestible compounds in feed and bovine feces obtained by *in situ* procedures. *Revista Brasileira de Zootecnia,* v.37, n.2, p.335-342, 2008.

CAVALCANTE, A.C.R.; PEREIRA, O.G.; VALADARES FILHO, S.C. et al. Diets containing corn silage (Zea mays L.) and Tifton 85 grass hay (Cynodon spp.) in different proportions for cattle. *Revista Brasileira de Zootecnia*, v.33, n.6, p.23942402, 2004.

COCHRAN, R. C.; ADAMS, D. C.; WALLACE J. D. et al. Predicting digestibility of different diets with internal markers: evaluation of four potential markers. *Journal of Animal Science*, v.63, p.1476-1483, 1986.

CONRAD, H. R., PRATT, A. D. and HIBBS, J. W. Regulation of feed intake in dairy cows. 1. Change in importance of physical and physiological factors with increasing digestibility. *Journal Dairy Science*, 47: 54-62. 1964.

COSTA, H. N.; PEREIRA, M. N.; MELO, R. P. et al. Effect of the rumen environment on ruminal in situ degradability of sugarcane. In: WORLD CONFERENCE ON ANIMAL PRODUCTION, 9. Porto Alegre: UFRGS, Anais... 2003.

COSTA, M. A. L.; VALADARES FILHO, S. C.; VALADARES, R. F. D. et al. Validation of the NRC (2001) equations for predicting the energy value of feeds under Brazilian conditions. *Revista Brasileira de Zootecnia,* v.34, p.280-287, 2005.

DETMANN, E.; PAULINO, M. F.; ZERVOUDAKIS, J. T. et al. Chromium and internal indicators in determining the intake of supplemented crossbred steers at pasture. *Revista Brasileira de Zootecnia*, v.30, p.1600-1609, 2001.

DETMANN, E.; SOUZA, A. L.; GARCIA, R.; et al. Evaluation of the "long time" bias of internal indicators in ruminant digestion assays. *Arquivo Brasileiro de Medicina Veterinària e Zootecnia*, v.59, n.1, p.182-188, 2007.

DIAS, M.; DETMANN, E.; LEAO, M. I. et al. Indicators for estimating partial digestibility in cattle. *Revista Brasileira de Zootecnia*, 36: 689-697. 2007.

DIAS, M. et al. Techniques for estimating digestibility and microbial production in cattle. *Revista Brasileira de Zootecnia*, Viçosa, v. 37, n. 6, p. 504-512, Jun, 2008.

FERREIRA, M. A.; VALADARES FILHO, S. C.; MARCONDES, M. I. et al. Evaluation of indicators in studies with ruminants: digestibility. *Revista Brasileira de Zootecnia*, v. 38, n. 8, p. 1568-1573, 2009.

FERREIRA, MA; VALADARES FILHO, SC; MARCONDES, MI et al. Evaluation of Indicators of Studies with RUMINANTS: Digestibility *Revista Brasileira de Zootecnia* , v.38, p.1568-1573, 2009a.

FREITAS, D.; BERCHIELLI, T. T.; SILVEIRA, R. N. et al. Fecal production and duodenal flow of dry matter and organic matter estimated by indicators. *Revista Brasileira de Zootecnia*, v.31, p.1521-1530, 2002.

GLINDEMANN, T.; TAS, B. M.; WANG, C. et al. Evaluation of titanium dioxide as an inert marker to estimate fecal excretion in grazing sheep. *Animal Science and Technology Feed* , v.152, p.186-197, 2009.

ÎTAVO, L. C. V.; VALADARES FILHO, S. C.; SILVA, F. F. et al. Intake, ruminal degradability and apparent digestibility of Cynodon grass hays and concentrate feeds using internal indicators. *Revista Brasileira de Zootecnia*, v.31, p.1024-1032, 2002.

LIMA, L. P.; VELOSO, C. M.; SILVA, F. F. et al. Cassava bagasse (*Manihot esculenta,* Crantz) in the diet of dairy cows: nutrient intake. *Arquivo Brasileiro Medicina Veterinâaria e Zootecnia,* v.60, p.1004-1010, 2008.

LIPPKE, H.; ELLIS, W. C.; JACOBS, B. F. Recovery of indigestible fiber from feces of sheep and cattle on forage diets. *Journal of Dairy Science*, v.69, n.2, p.403-412, 1986.

MARCONDES, M.I.; VALADARES FILHO, S.C.; BRITO, A.F. et al. Use of different indicators to estimate fecal dry matter production and to evaluate the individual consumption of concentrate and roughage in heifers. In: ANNUAL MEETING OF THE BRAZILIAN SOCIETY OF ZOOTECHNICS, 43. Joao Pessoa: SBZ, Anais... 2006. p.1-5.

MAYES, R.W. The quantification of dietary intake, digestion and metabolism in farm livestock and its relevance to the study of radionuclide uptake. *Science Total Environ*, v.85, p.29-51, 1989.

MENDONÇA, S. S.; CAMPOS, J. M. S.; VALADARES FILHO, S. C. et al. Ingestive behavior of dairy cows fed diets based on sugarcane or corn silage. *Revista Brasileira de Zootecnia*, v.33, p.723-728, 2004.

MERCHEN, N. R. Digestion, absorption and excretion in ruminants. In: CHURCH, D.C. (Ed.) The ruminant animaldigestive physiology and nutrition. Englewood Cliffs: Prentice Hall, 1988. p.172-201.

MERLO, F. A.; SILVA, A. G. M.; BORGES, I. et al. Evaluation of LIPE as an external indicator of digestibility in sheep fed diets containing coconut bran. In: ZOOTEC. Joao Pessoa: ABZ, Anais. 2008.

MERTENS, D. R. Regulation of forage intake. In: FAHEY Jr., G.C. Forage quality, evaluation, and utilization. Lincoln: University of Nebraska, 1994. p.450-493.

MORAES, K. A. K.; VALADARES FILHO, S. C.; MORAES, E. H. B. K. et al. Consumption of beef heifers fed sugar cane with calcium oxide and different concentrate offerings. In: ANNUAL MEETING OF THE BRAZILIAN SOCIETY OF ZOOTECHNICS, 44, Jaboticabal: Anais... SBZ, 2007. CD-ROM.

MYERS, W. D.; LUDDEN, P. A.; NAYIGIHUGU, V. et al. Technical Note: a procedure for the preparation and quantitative analysis of samples for titanium dioxide. *Journal of Animal Science*, v.82, n.1, p.179-183, 2004.

MYERS, W. D.; LUDDEN, P. A; NAYGIHUGU, V. et al. Excretion patterns of titanium dioxide and chromium oxide in duodenal digesta and feces of sheep. *Pesquisa de Pequenos Ruminantes* , v.63, p.135-141, 2006.

NATIONAL RESEARCH COUNCIL. Nutrient requirements of dairy cattle. 7.ed. Washington, DC: *National Academy Press*, 2001. 381p.

OLIVEIRA JUNIOR, R. C.; PIRES, A. V.; FERNANDEZ, J. J. R. et al. Evaluation of indicators to estimate nutrient digestibility in Nelore steers fed diets containing high concentrate content and nitrogen sources. *Revista Brasileira de Zootecnia*, v.33, n.3, p.749-758, 2004.

OLIVEIRA, M. A. Proportion of forage and lipid content in the diet of dairy cows on consumption, production and milk composition. 2005. 63f. Dissertation (Master's Degree in Animal Science) -

Federal University of Minas Gerais, Veterinary School, Belo Horizonte.

OWENS, F. N.; HANSON, C. F. External and internal markers for appraising site and extent of digestion in ruminants. *Journal of Dairy Science*, v. 75, n. 9, p. 2605-2617, 1992.

PAIXÂO, M. L.; VALADARES FILHO, S. C.; LEÂO, M. I. et al. Daily variation in the excretion of internal (FDAi) and external (Cr 2O3) indicators, digestibility and ruminal parameters in cattle fed diets containing urea or soybean meal. *Revista Brasileira de Zootecnia,* v.36, n.3, p. 739-747, 2007.

PENNING, P. D.; JOHNSON, R. H. The use of internal markers to estimate herbage digestibility and intake. Journal of Agricultural Science, v.100, p.127-131, 1983.

PERIPOLLI, V. Use of fecal nitrogen to estimate intake and digestibility in grazing ruminants. Porto Alegre: Faculty of Agronomy, Federal University of Rio Grande do Sul, 2010, 76p. Dissertation (Master's Degree in Animal Science).

PIAGGIO, L. M.; PRATES, E. R.; PIRES, F. F. et al. Evaluation of acid-insoluble ash, indigestible acid detergent fiber and indigestible acid detergent lignin as internal indicators of digestibility. *Revista da Sociedade Brasileira de Zootecnia*, v.20, n.3, p.306-312, 1991.

PINA, D.S. Nutritional evaluation of sugar cane plus calcium oxide at different storage times for cattle. 104f. 2008. Thesis (Doctorate in Animal Science) - Federal University of Viçosa, Viçosa, MG.

RANGEL, A.H.N; CAMPOS, J.M.S; OLIVEIRA, A.S. et al.; In: Performance and nutritional parameters of growing dairy females fed corn silage or sugarcane with concentrate. *Revista Brasileira de Zootecnia*, v.39, n.11, p.2518-2526, 2010.

RODRIGUEZ, N. M.; GUIMARAES JUNIOR, R. Utilization of agro-industry by-products in the feeding of dairy cows. In: SIMPÒSIO MINEIRO DE NUTRIÇÂO DE GADO DE LEITE, 3. Belo Horizonte: EV/UFMG, Anais... 2005. p.65- 91.

RODRIGUEZ, N. M; SALIBA, E. O. S.; GUIMARAES JÙNIOR, R. Use of indicators to estimate pasture intake and digestibility. In: 43ª REUNIAO ANUAL DA SOCIEDADE BRASILEIRA DE ZOOTECNIA, Joao Pessoa. SBZ, Anais...2006.

SALIBA, E. O. S. Chemical and microscopic characterization of lignins from corn and soybean residues exposed to rumen degradation and their effect on the digestibility of structural carbohydrates. Belo Horizonte: Veterinary School of the Federal University of Minas Gerais, 1998, 251p. Thesis (Doctorate in Animal Science).

SALIBA, E. O. S.; RODRIGUEZ, N. M.; PILO-VELOSO, D. Purified lignin extruded from

Eucalyptus grandis (PELI), used as an external marker in digestibility trials in various animal species. In: WORLD CONFERENCE ON ANIMAL PRODUCTION, 9, Porto Alegre: WAAP/ALPA/SBZ/UFRGS; Anais... 2003.

SAMPAIO, c. B.; DETMANN, E.; VALENTE, T. N. P. et al. Evaluation of fecal recovering and long term bias of internal and external markers in a digestion assay with cattle. *Revista Brasileira de Zootecnia*, v.40, p.174-182, 2011.

SILVA, A. G. M.; BORGES, I.; NEIVA, J. N. et al. Evaluation of LIPE as an external indicator of digestibility in sheep receiving diets with whole cashew nuts 1. IN: ZOOTEc, Joao Pessoa: ABZ, Anais... 2008.

SILVA, A. G. M.; BORGES, I.; NEIVA, J. N. et al. Evaluation of LIPE as an external indicator of digestibility in sheep fed diets containing babassu cake. IN: I BRAZILIAN CONGRESS ON ANIMAL NUTRITION, Fortaleza: AMVEcE, Anais 2008b.

SILVA, J. J.; SALIBA, E. O. S.; BORGES, I. et al. Indicators for estimating total intake by Holstein x Zebu heifers kept in confinement. *Revista Brasileira Saùde e Produçào Animal*, v. 11, n. 3, p 838-848, 2010.

SILVA, F. A.; BARBOSA, G. S. S. c.; SALIBA, E. O. S. et al. Evaluation of LIPE® in estimating the consumption of dairy cattle. In: 48th Annual Meeting of the Brazilian Society of Animal Science, 2011, Belém. ***Anais...*** da 48a Reuniao da Sociedade Brasileira de Zootecnia. Belém: RBZ, 2011.

SNIFFEN, C. J.; O'CONNOR, J. D.; VAN SOEST, P. J. et al. A net carbohydrate and protein system for evaluating cattle diets: II. Carbohydrate and protein availability. *Journal Animal Science*, v.70, p.3562-3577, 1992.

TITGEMEYER, E. C. Design and interpretation of nutrient digestion studies. *Journal Animal Science*, v. 75, p. 2235-2247, 1997.

TITGEMEYER, E. C.; AMENDARIZ, C. K.; BINDEL, D. J. et al. Evaluation of titanium dioxide as a digestibility marker in cattle. *Journal Animal Science.* 79:1059-1063. 2001.

TILLEY, J.M.A.; TERRY, R.A. A two-stage technique for the "in vitro" digestion of forage crops. *Journal of British Grassland Society***.** v.18, n.2, p.104-111, 1963.

TORRES, L. C. L. et al. Substitution of giant palm for half palm in diets for growing cattle and evaluation of internal indicators. *Revista Brasileira de Zootecnia***.** vol. 38 n° .11 Viçosa Nov. 2009.

TRAXLER, M.J.; FOX, D.G.; PERRY, T.C. et al. Influence of roughage and grain processing in high-concentrate diets on the performance of long-fed Holstein steers. *Journal of Animal Science*, v.73, p.1888-1900, 1995

VALADARES FILHO, S. C. Nutrition, feed evaluation and feed composition tables for cattle. In: Annual Meeting of the Brazilian Society of Animal Science, Viçosa, MG. ***Anais...*** Sociedade Brasileira de Zootecnia, p.267-337. 2000.

VALADARES FILHO, S. C.; MAGALHAES, K. A.; ROCHA Jr.; V. R. et al. Tabelas brasileiras de composiçao de alimentos para bovinos. 2.ed. Viçosa: DZO/UFV, 2006. 329p.

VAN SOEST, P. J. Nutritional ecology of the ruminant. New York: Cornell University Press, 1994, 476p.

ZEOULA, L. M.; PRADO, I. N.; DIAN, P. H. M. et al. Fecal recovery of internal markers evaluated in ruminants. *Revista Brasileira de Zootecnia*, v.31, n.4, p.1865- 874. 2002.

MIX
Papier aus verantwortungsvollen Quellen
Paper from responsible sources
FSC® C105338

Printed by Books on Demand GmbH, Norderstedt / Germany